ISBN 978-0-260-66198-2
PIBN 11118091

AIC-321
Supplement 2

SEMIANNUAL LIST OF PUBLICATIONS AND PATENTS WITH ABSTRACTS

Western Regional Research Laboratory
Albany 6, California

JANUARY 1 TO JUNE 30, 1952

CONTENTS:

Lists with abstracts of recent publications and patents of other units of the Bureau of Agricultural and Industrial Chemistry in the Western Region are included as follows:

Enzyme Investigations
Pharmacological Investigations
Natural Rubber Evaluation and Processing
(Salinas, Calif.)
Fruit and Vegetable Chemistry Laboratory
(Pasadena, Calif.)

Publications with AIC preceding their numbers (for Example AIC-337) are mimeographed, and supplies are maintained for free distribution. A limited number of reprints of journal articles are available, usually for a short period following publication. The asterisk (*) before a title indicates that the supply of reprints of that publication was exhausted at the time this list was prepared. Photostat copies can be purchased at nominal cost through the Library of the United States Department of Agriculture, Washington 25, D. C. All patents are assigned to the U. S. Government. Copies of patents must be purchased from the United States Patent Office, Washington, D. C., for 25 cents each.

Bureau of Agricultural and Industrial Chemistry
Agricultural Research Administration
UNITED STATES DEPARTMENT OF AGRICULTURE

ALFALFA

PYROGALLOL DERIVATIVES AS ANTIOXIDANTS FOR CAROTENE. E. M. Bickoff, G.M. Coppinger A. L. Livingston, and T. W. Campbell, Jour. Amer. Chem. Soc. 29(2):51-53, Feb., 1952. Certain derivatives of pyrogallol have been shown to possess marked activity as antioxidants for carotene in some systems, while they are of little value in others.

EFFECT OF HEAT BLANCHING ON ALFALFA. C. R. Thompson, Agri. Engin. 33(1):1920, Jan., 1952. (The work reported was conducted when the author was employed at the University of Chicago Botany Laboratory, Riverside, Calif. It is listed here because reprints are available at the Western Regional Research Laboratory, where the author is now employed.) The studies showed that heating green alfalfa as soon as it is cut in the field hastens drying and preserves carotene. A design for a machine that would accomplish this operation is proposed.

ANALYTICAL AND PHYSICAL

(Analytical papers concerned with specific subjects are listed elsewhere.)

COLOR REACTION OF SOME ALDEHYDES WITH THE ORCINOL REAGENT AT DIFFERENT TEMPERATURES A. Bevenue and K. T. Williams, Chem. Analyst. 41(1):5-7, March, 1952. A group of aromatic aldehydes were found to produce color when treated with an orcinol reagent on filter paper. Cinnamaldehyde, p-dimethylaminobenzaldehyde, and p-hydroxybenzaldehyde gave characteristic color reactions.

THE OLIGOSACCHARIDES FORMED DURING THE SUCROSE-INVERTASE REACTION. L. M. White and G. E. Secor, Arch. Biochem. and Biophys. 36(2):490-91, April, 1952. Chromatographi techniques developed in this Laboratory and elsehwere were used to study the oligosaccharides formed when sucrose is incompletely hydrolyzed by invertase. Ratios of fructose to glucose units in the saccharides formed were quantitatively determined. These ratios and some of the chromatographic properties of the sugars are given.

A STUDY, BY PAPER CHROMATOGRAPHY, OF THE OCCURRENCE OF NONFERMENTABLE SUGARS IN PLANT MATERIALS. K. T. Williams, E. F. Potter, and A. Bevenue, Jour. Assoc. Offic. Agri. Chem. 35(2):483-86, May, 1952. Ketoheptoses are present in a large variety o plant materials. Because of the frequent occurrence of various nonfermentable sugars, a preliminary investigation by means of paper chromatography would aid in proper interpretation of sugar values obtained by methods of chemical analysis.

REPORT ON SUGARS IN PLANTS. K. T. Williams, Jour. Assoc. Offic. Agr. Chem. 35(2): 402, May, 1952. In a continued study of clarification in the analysis of leaf extracts, the ion exchange resin technique was equal or superior to the lead-acetat technique.

D-FRUCTOSE DIHYDRATE. F. E. Young, F. T. Jones, and H. J. Lewis, Jour. Phys. Chem. 36(6):738-39, June, 1952. The discovery, preparation, and properties of a new crystalline di-hydrate of B-D-fructose are described.

METHOD AND APPARATUS FOR HEATING FLUIDS. Patent No. 2,585,970 to T. M. Shaw, Feb. 19, 1952. A thin stream of fluid is caused to fall through a cylindrical cavity on a path coincidental with the axis of the cavity. A generator of high-frequency electromotive waves (500-5000 megacycles per second) is connected with the cavity to establish within it an electric field which heats the stream to a uniform and maximum extent because the field is at its maximum along the path of the stream.

LEVULOSE DIHYDRATE. Patent No. 2,588,449 to F. E. Young and F. T. Jones, March 11, 1952. Levulose is isolated as its dihydrate from solutions containing levulose. An aqueous solution containing in excess of 36 percent levulose and not over 18 percent dextrose is cooled (-20°C.) and seeded with crystals of levulose dihydrate. The levulose dihydrate crystallizes out and is removed from the mother liquor. The process can be used for isolating levulose, valuable because of its high sweetening power, from sugar mixtures such as invert sugar.

METHOD AND APPARATUS FOR MEASURING DEW POINT. Patent No. 2,588,355, to H. K. Burr and G. F. Bailey, March 11, 1952. Air to be tested is blown over a mirror which is warm at one end and cool at the other. The mirror will become fogged to an extent dependent on the dew point of the air, the fogged zone extending from the cool end of the mirror toward the warm end and ending abruptly at the point on the mirror where the mirror temperature and dew point coincide. By measurement of the length of the fogged zone and the temperature at each end of the mirror, the dew point of the air can be rapidly calculated.

ANTIBIOTICS

SUBTILIN CONSIDERED AS A GERMICIDAL SURFACE-ACTIVE AGENT. L. E. Sacks, Antibiotics and Chemotherapy 2(2):79-85, Feb., 1952. Subtilin causes rapid and permanent decrease in metabolic activity of *Micrococcus pyogenes* var. *aureus* and *M. conglomeratus*, rapid decrease in viable count, and leakage of phosphorus and nitrogen compounds from the cells. It retains its activity in lecithin and cephalin preparations.

THE FREE AMINO GROUPS OF SUBTILIN. J. F. Carson, Jour. Amer. Chem. Soc. 74(6): 1480-82, March, 1952. Deamination of the polypeptide, subtilin, with nitrous acid, followed by hydrolysis and microbiological assay and reaction of subtilin with dinitrofluorobenzene, followed by hydrolysis and chromatographic examination of the DNP amino acids, show that the free amino groups of subtilin are contributed by lysine and the two sulfur diamino dicarboxylic acids.

FRUIT

RAPID PREPARATION OF THICK SECTIONS OF FLESHY PLANT TISSUES. R. M. Reeve, Stain Technol. 27(1):29-35, Jan., 1952. Methods are described for permanent microslide preparations of soft, large-celled plant tissues such as ripe fruit.

INFLUENCE OF STORAGE CONDITIONS ON FROZEN AND CANNED APPLE JUICE CONCENTRATES. C. C. Nimmo, L. H. Walker, and V. S. Seamans, Food Technol. 6(1):31-35, Jan., 1952. Apple juice concentrates prepared by three different methods have been stored at 0°F. and 75°F. and the changes in flavor, color, and iron concentration have been followed. Storage at 0°F. for one year produced no important flavor changes. Storage at 75°F. for more than three months caused development of an objectionable amount of off flavor.

ADVANCED FRUIT WASTE RECOVERY WILL ROLL DOLLARS IN--NOT OUT. R. P. Graham, A. D. Shepherd, A. H. Brown, and W. D. Ramage, Food Engin. 24(2):82-83, 151-53, Feb., 1952. A fruit-cannery waste material is converted to salable molasses and dried pulp for stock feed and other uses. Utilization of the fruit waste promises to eliminate disposal costs and yield a profit.

NEW APPLE SHERBET FLAVOR. D. G. Guadagni, L. H. Walker, W. F. Talburt, and R. Farris (with Borden Dairy Delivery Co.), Ice Cream Field 59(3):26, March, 1952. Diced fresh apple impregnated with concentrated apple juice is incorporated into sherbet. The juice has been concentrated after removal of essence, which is also concentrated and returned to the juice. This method gives a true apple flavor to sherbet.

ENZYMATIC BROWNING OF FRUITS. III. KINETICS OF THE REACTION INACTIVATION OF POLYPHENOLOXIDASE. L. L. Ingraham, J. Corse, and B. Makower, Jour. Amer. Chem. Soc. 74(10):2623-26, May, 1952. In the enzymatic oxidation of catechol by apple enzyme in the presence of an excess of ascorbic acid the amount of catechol oxidized (as measured by the ascorbic acid destroyed by the o-quinone or semiquinone produced) is proportional to the amount of enzyme added.

CORRELATION OF O-H STRETCHING FREQUENCIES IN PHENOLS AND CATECHOLS WITH CHEMICAL REACTIVITIES. L. L. Ingraham, J. Corse, G. F. Bailey, and F. Stitt, Jour. Amer. Chem. Soc. 74(9):2297-99, May, 1952. The O-H stretching frequencies have been studied for a series of substituted catechols and phenols. The frequency shifts produced by substitution have been related to Hammett's sigma values.

*THE REACTION INACTIVATION OF POLYPHENOLOXIDASE. L. L. Ingraham, Vortex 13(5):220, 222, May, 1952. A review of work at Western Regional Research Laboratory on polyphenoloxidase.

PECTIN

AIC-340, METHODS USED AT WESTERN REGIONAL RESEARCH LABORATORY FOR EXTRACTION AND ANALYSIS OF PECTIC MATERIALS. H. S. Owens, R. M. McCready, A. D. Shepherd, T. H. Schultz, E. L. Pippen, H. A. Swenson, J. C. Miers, R. F. Erlandsen, and W. D. Maclay June, 1952. Methods used in the Western Regional Research Laboratory for the extraction, isolation, purification, and analysis of pectic substances are described. Later pages contain references and a list of current nomenclature of pectic substances and enzymes.

PREPARATION AND STORAGE OF AUTOCLAVED PECTIN SOLUTIONS. T. H. Schultz, H. Lotzkar, H. S. Owens, and W. D. Maclay, Jour. Amer. Pharm. Assoc. 41(5):251-57, May, 1952. Studies on degraded pectin solutions, designed for possible use as blood-plasma extenders, include filtration rates, type of filter, and rate of change under various storage temperatures.

ACYLATION OF POLYSACCHARIDES IN FORMAMIDE. Patent No. 2,589,226 to J. F. Carson, March 18, 1952. A polysaccharide such as starch, pectin, pectic acid, etc., is dispersed in formamide and then reacted with an acyl anhydride. Use of the formamide simplifies the esterification as it eliminates various pre-treatment steps formerly used to gelatinize the polysaccharide to make it reactive.

POULTRY

EFFECT OF ANTIOXIDANT ON RANCIDITY DEVELOPMENT IN FROZEN CREAMED TURKEY. H. Lineweaver, J. H. Anderson, and H. L. Hanson, Food Technol. 6(1):1-4, Jan., 1952. Presence of an antioxidant during cooking of turkey to be used in precooked frozen foods had a beneficial effect in preventing development of rancid flavors and peroxide both during cooking and during subsequent frozen storage. Addition of antioxidant during preparation of the cream sauce rather than during cooking of the turkey further retarded development of rancid flavors and peroxide.

RELATIVE EFFICIENCY OF DIFFERENT CULTURE MEDIUMS IN ISOLATION OF CERTAIN MEMBERS OF THE SALMONELLA GROUP. N. B. McCullough and A. F. Byrne. (University of Chicago, research conducted under contract.) Jour. Infect. Diseases 90(1):71-75, Jan.-Feb. 1952. Data showing relative efficiency of a variety of media in detecting certain strains of Salmonella in presumptively positive fecal specimens is presented. The species of Salmonella used were: S. meleagridis, S. anatum, S. newport, S. bareilly, S. derby, and S. pullorum.

CHEMICAL CHARACTERISTICS OF TURKEY CARCASS FAT AS A FUNCTION OF DIETARY FAT. A. A. Klose, E. P. Mecchi, G. A. Behman, and H. Lineweaver (WRRL) and F. H. Kratzer and D. Williams (Univ. California). Poultry Sci. 31(2):354-59, March, 1952. Composition of turkey carcass fat has been established as a function of type of dietary fat present. The relatively greater stability of depot fat synthesized by the turkey and the much less stability of depot fat of birds fed unsaturated fatty acids both indicate the importance of scrutinizing the fat components of commercial diets.

INACTIVATION OF LYSOZYME BY TRACE ELEMENTS. R. E. Feeney, E. D. Ducay, and L. R. MacDonnell. Fed. Proc. 11(1):pt. I, 209-210, March, 1952. Lysozyme was found extremely stable in alkaline solution in the absence of trace metal ions.

PROCESS FOR THE TREATMENT OF EGG WHITES. Patent No. 2,585,015 to A. M. Kaplan and M. Solowey, Feb. 12, 1952. Prior to dehydration, egg whites are fermented at 21-37°C., pH 6.7-9.0, with a concentrated preparation of Streptococcus lactis in order to rapidly assimilate the reducing sugars naturally present in egg white and thus produce a dried product of improved odor and flavor.

METHOD OF TREATING EGG WHITES. Patent No. 2,593,463 to A. M. Kaplan and M. Solowey, April 22, 1952. Prior to dehydration, egg whites are fermented with a concentrated preparation of a micro-organism selected from the genera Lactobacilli, Streptococci, and Aerobacter in order to rapidly assimilate the reducing sugars naturally present in egg white and thus produce a dried product of improved odor and flavor.

SUGAR BEETS

THE HYDROLYSIS RATE OF BETAINE AMIDE AND ESTER CHLORIDES. H. G. Walker, Jr. and H. S. Owens, Jour. Amer. Chem. Soc. 74(10):2547-49, May, 1952. The acidic and alkaline hydrolysis rates of a typical betaine ester and betaine amide have been studied. These derivatives are hydrolyzed faster in alkali and slower in acid than similar uncharged compounds.

CONSERVATION OF SUGAR BEET TOPS BY DEHYDRATION. R. D. Barmington, P. N. Davis, and H. S. Wilgus, Tech. Bull. 47, Colorado Agri. Expt. Sta., May, 1952. (The research on dehydration was conducted under a contract supervised by Western Regiona Research Laboratory.) This bulletin reports studies on nutritive components of beet tops, effects of fertilization, mechanical harvest, dehydration, and economics of top utilization.

VEGETABLES

HYGROSCOPIC EQUILIBRIUM OF BROWN RICE. D. F. Houston, Cereal Chem. 29(1):71-76, Jan., 1952. Equilibrium moisture contents of Caloro brown rice at 25°C. have been determined for atmospheric relative humidities of 11 to 93 percent, together with rates of approach to equilibrium. Extent of hysteresis effects in sorption of moisture is indicated by the fact that rice brought to equilibrium at 40 to 65 percent relative humidity from a dry condition attains a moisture content 0.7 to 1.0 percent lower than that resulting from direct drying of moist rice to the same relative humidity.

FROZEN COOKED RICE. II. BROWN RICE. M. M. Boggs, A. C. Ward, C. N. Sinnott, and E. B. Kester, Food Technol. 6(2):53-54, Feb., 1952. Samples of brown rice were cooked by approximate methods, frozen, stored at 0°F., and tasted after a year of storage. The quality of the frozen product was equivalent to that of freshly prepared samples.

EFFECT OF BRAN DAMAGE ON DEVELOPMENT OF FREE FATTY ACIDS DURING STORAGE OF BROWN RICE. D. F. Houston, E. A. McComb, and E. B. Kester, Rice Jour. 55(2):17-18, Feb., 1952. Brown rice prepared with rubber shelling rolls has an almost undamaged bran layer, as compared to brown rice dehulled with the usual shelling stones. Prevention of bran damage is clearly shown to increase storage qualities of brown rice.

RAPID METHOD FOR MOISTURE IN FRUITS AND VEGETABLES BY OXIDATION WITH DICHROMATE. I. POTATOES AND PEAS. H. F. Launer and Y. Tomimatsu, Food Technol. 6(2):59-64, Feb., 1952. A rapid, simple method, designed for measurement of moisture in fresh or moist materials but which may also have limited applicability to dry materials, is based upon oxidation with potassium dichromate solution. Data are presented on precision and accuracy for dehydrated and fresh materials.

REACTION OF T-BUTYLHYDROPEROXIDE WITH SOME PHENOLS. T. W. Campbell and G. M. Coppinger, Jour. Amer. Chem. Soc. 74(6):1469-71, March, 1952. The phenolic antioxidant 2,6-di-t-butyl-p-cresol is converted by reaction with t-butylhydroperoxide to 1-methyl-1-t-butylperoxy-3-5-di-t-butylchclohexadiene-2,5-one-4. This peroxide reacts with bromine in acetic acid to give in high yield 3,5-di-t-butyl-4-hydroxybenzaldehyde.

DEVELOPMENT OF OFF FLAVOR IN SHELLED PEAS HELD IN COLD WATER. R. U. Makower, M.M. Boggs, and H. K. Burr, Food Technol. 6(5):179-80, May, 1952. Shelled Giant Stride peas held dry or submerged in water at different temperatures before blanching and freezing were organoleptically inferior to controls processed immediately after shelling. Off-flavor was stronger in submerged samples than in the dry. In general the lower the temperature of holding, the less the off-flavor produced.

RAPID DETERMINATION OF MOISTURE IN DEHYDRATED VEGETABLES WITH KARL FISCHER REAGENT BY USE OF FORMAMIDE AS AN EXTRACTION SOLVENT. E. A. McComb and R. M. McCready, Jour. Assoc. Offic. Agr. Chem. 35(2):437-41, May, 1952. Use of formamide as extraction solvent has made possible the rapid determination of moisture in dehydrated white potatoes, sweetpotatoes, carrots, and peas by the Karl Fischer titration method. Values obtained are in good agreement with a vacuum oven reference method.

ISOLATION AND CHARACTERIZATION OF TRYPSIN INHIBITOR FROM LIMA BEANS. H. Fraenkel-Conrat, R. C. Bean, E. D. Ducay, and H. S. Olcott, Arch. Biochem. and Biophys. 37(2):393-407, June, 1952. The purification of a lima bean protein fraction of high trypsin-inhibiting activity is described. The active protein was freed from inert crystallizable protein.

PREPARATION OF NUT-LIKE PRODUCTS FROM DRY BEANS. Patent No. 2,584,150 to H. J. Morris, Feb. 5, 1952. Dry beans (Michelite, Great Northern, Red Mexican, etc.) are soaked in water, partially cooked in water, fried in oil, and roasted until crisp. The products are suited for out-of-hand eating and have a crisp texture and a distinctive nutty flavor.

PREPARATION OF NOVEL FOOD PRODUCTS FROM RICE. Patent No. 2,585,036 to R. L. Roberts, Feb. 12, 1952. Rice meal (16-24 mesh) is mixed with hot water to form a paste which is extruded into elongated shapes, preferably rods about 3/16 inch diameter. These are fried in hot oil, 375-425°F., until brown, cooled, and sprinkled with salt to produce a tasty product for use as an hors d'oeuvre or as a supplement to beverages and soups.

METHOD OF STABILIZING RICE PRODUCTS. Patent No. 2,585,978 to G. R. Van Atta, E. B. Kester, and H. S. Olcott, Feb. 19, 1952. Rough (unhulled or paddy) rice is subjected to a moist heat treatment (185-212°F., 0.5 to 15 min.) to inactivate the lipase present in the grain. The treated rough rice, when milled, yields rice bran and/or brown rice which is stabilized in that it can be stored for long periods without developing rancidity.

WOOL

THE MOLECULAR SIZE OF PROTEINS FROM SEVERAL WOOLS SOLUBILIZED IN AQUEOUS UREA. W. H. Ward, Textile Res. Jour. 22(6):405-15, June, 1952. The average molecular weight from sedimentation and diffusion in 10M urea is near 13,000. The high molar fractional ratio, 2.8, suggests high solvation, high asymmetry, or both. Normal wools differing widely in geographical origin or in fineness from different breeds of sheep show little variation in diffusion or sedimentation. The differences in properties of normal wools may be due to differences in proportion of soluble and insoluble components.

ACROLEIN AND SULFUR HALIDE PROCESS TO STRENGTHEN PROTEIN FIBERS. Patent No. 2,583,574 to H. W. Jones, W. J. Thorsen, and H. P. Lundgren, Jan. 29, 1952. Natural protein fibers such as wool or protein fibers made from feathers, casein, zein, etc., are impregnated with acrolein and then treated with a sulfur halide such as sulfur monochloride to strengthen the fiber.

PROPERTIES OF APPAREL WOOLS. I. PRELIMINARY REPORT ON WORSTED PROCESSING TRIALS. W. von Bergen and J. H. Wakelin, Forstmann Woolen Co. and Textile Research Institute (research conducted under contract supervised by Western Regional Research Laboratory). Textile Research Jour. 22(2):123-37, Feb., 1952. Studies have been made on the physical and chemical properties and worsted processing characteristics of two fine and two medium wools. Each of the wools studied was found to have essentially the same basic physical properties when reduced to the same cross-sectional area and to have the same chemical composition. The average diameter or fineness of a wool lot is the most important characteristic of the wool insofar as the efficiency of manufacturing operation is concerned. By aging the wool at 70°F. and 70 percent R.H., an improvement in spinning and weaving was observed. Steaming the wool in the form of roving was found to have the same beneficial effect.

RESEARCH ACHIEVEMENT SHEETS

The Agricultural Research Administration of the U. S. Department of Agriculture is responsible for the issuance of Research Achievement Sheets, each of which describes briefly an achievement in research in agriculture and also presents estimates of costs of the research, estimates of values, and other data related to the achievement. The following sheets are concerned with work of the Bureau of Agricultural and Industrial Chemistry in the western region and are available on request.

90(C), Techniques of Vegetable, Fruit, and Egg Dehydration Advanced.

116(C), New Froth Flotation Process Cleans Vined Peas.

150(C), Orange Maturity Standard Unchallenged After Test of Time.

151(C), Hull-Loosening Process Improves Walnut Kernels.

152(C), Ethylene Coloring of Citrus Successful Over Long Period.

ENZYME RESEARCH, July, 1950 to June 30, 1952. (A separate list, AIC-219, with supplements I-III, includes abstracts of publications and patents of the Enzyme Research Division of this Bureau prior to July, 1950.)

BETA-AMYLASE FROM SWEET POTATOES. Patent No. 2,496,261 to A. K. Balls, M. K. Walden, and R. T. Thompson, Feb. 7, 1950. Pure crystalline beta-amylase is obtained from sweet potatoes by clarifying the press juice with $Pb(AcO)_2$, precipitating the enzyme and other protein by partial saturation with $(NH_4)_2SO_4$, dialyzing the precipitate against water to remove the sulfate, acidifying the dialyzate in a number of stages successively to lower pH, and removing the impurities precipitated at each stage. The solution is then partially saturated with $(NH_4)_2SO_4$ to precipitate impurities, increase the $(NH_4)_2SO_4$ content, and to precipitate beta-amylase.

PREPARATION OF PEPTIDES. Patent No. 2,498,665 to O. H. Emerson, Feb. 28, 1950. Peptides are prepared by reacting a phthalyl amino acid halide with an amino acid and then refluxing the phthalyl peptide so formed with hydrazine to obtain the peptide.

EFFECT OF HIGH PRESSURES ON ACTIVITY OF PROTEINASES. A. L. Curl and E. F. Jansen, Fed. Proc. 9, No. 1, part 1, 164, March, 1950. The effect of high pressures on activation of chymotrypsinogen by trypsin has been studied in connection with a general investigation of the effect of high pressures on enxymes, in particular crystalline proteinases.

EFFECT OF HIGH PRESSURES ON PEPSIN AND CHYMOTRYPSINOGEN. A. L. Curl and E. F. Jansen, Jour. Biol. Chem. 185(2):71323, Aug., 1950. The effect of high pressures on pepsin and chymotrypsinogen were investigated, including the influence of pH, concentration, magnitude of pressure, duration of pressing, and multiple pressing. Results with pepsin differed in several important respects from those previously obtained with trypsin and chymotrypsin.

KINETICS OF MALT ALPHA-AMYLASE ACTION. S. Schwimmer, Jour. Biol. Chem. 186(1): 181-93, Sept., 1950. Amylose is hydrolyzed about twice as fast as amylopectin. The enzyme has the same affinity for all substrates whose chain length averages more than 10-15 glucose units. End products of the reaction inhibit non-competitively, whereas intermediate products inhibit competitively. Fifty percent of the total glycosidic linkages are ultimately hydrolyzed.

ON THE "DEXTRINASE" ACTIVITY OF CRYSTALLINE AMYLASE PREPARATIONS. S. Schwimmer, Cereal Chem. 28(1):77-78, Jan., 1951. Malt extract, but not crystalline malt alpha-amylase, shows true glycosidase activity toward a dextrin 8 glucose units long. Beta-amylase has practically no action.

AMYLASE OF BACILLUS MACERANS. W. S. Hale and L. C. Rawlins, Cereal Chem. 28(1): 49-58, Jan., 1951. A scheme for growing Bacillus macerans on a laboratory scale larger than usual has been devised, and with its help an 8-fold purification (with respect to protein) of the macerans amylase in filtered culture fluid has been accomplished. The most effective step in purification has been adsorption of the enzyme on raw starch. Properties of the purified material indicate that a single enzyme is effective in producing Schardinger dextrins not only from starch but also from intermediate products of ordinary amylolysis. Its rate of action on starch is decreased by beta-Schardinger dextrin and accelerated by glucose or maltose.

3,3'-DITHIOLISOBUTYRIC ACID. Patent No. 2,539,428 to E. F. Jansen, Jan. 30, 1951. Several methods for preparing 3,3'-dithiolisobutyric acid are disclosed, for example by reacting 3,3'-diiodoisobutyric acid with sodium hydrosulphite. The product is useful as an intermediate for organic syntheses and as a regulator for enzymic reactions.

PLANT LIPASES IN EMULSIONS OF WATER IN OIL. W. G. Rose, Jour. Amer. Oil Chem. Soc. 28, 47-51, Feb., 1951. A procedure for preparing an active lipase cream of reproducible activity is described. Beans are ground in a mortar to a greasy paste, water being incorporated by frequent regrinding. After standing over night at 2-4° the pH is adjusted to 4.5, the suspension is centrifuged and the cream washed twice with water. The rate of splitting of cottonseed oil at -16 to -18°, 2 to 4°, 10°, 20°, and 25° is reported. The lipase is inhibited by soluble synthetic cephalin, egg phosphatide and salmon egg phosphatide, but not by soybean phosphatide.

ENZYMICALLY INDUCED CHANGES IN THE TURBIDITY OF STARCH SOLUTIONS. S. Schwimmer, Jour. Biol. Chem. 188(2):477-84, Feb., 1951. Alpha-amylase, beta-amylase, and B. macerans amylase each gives a typical turbidity pattern. Alpha-amylase and crude macerans amylase induce an initial increase in turbidity of starch solutions, followed by a decrease. Beta-amylase and purified macerans amylase cause a smooth decrease in turbidity to a constant value.

THE FATE OF PHOSPHATASE, INVERTASE, AND PEROXIDASE IN AUTOLYZING LEAVES. B. Axelrod and A. T. Jagendort (with Calif. Inst. of Technol.), Plant Physiol 26(2):406-10, April, 1951. The following enzymes were selected for study in the present work: phosphatase because of our special interest in its metabolic function, invertase and peroxidase because of the ease with which they could be determined. Loss of protein nitrogen in the cytoplasm of stored tobacco leaves is not reflected in a marked change in the contents of phosphatase, invertase, and peroxidase even when loss of protein amounts to 45 percent of that originally present.

THE MALT AMYLASES, A REVIEW OF RECENT LITERATURE. S. Schwimmer, Brewers Digest 26(3-4):29-32T, 40T, 43-45T, 48T, March-April, 1951. The present status of our knowledge of the enzymes including determination, crystallization, general properties, and action on amylose and amylopectin.

REACTION OF ALPHA CHYMOTRYPSIN WITH ANALOGUES OF DIISOPROPYL FLUOROPHOSPHATE (DFP). E. F. Jansen, A. L. Curl, and A. K. Balls, Fed. Proc. 10(1)pt. I, p. 203, March, 1951; Jour. Biol. Chem. 190(2):557-61, June, 1951. Analyses of the crystalline, inert proteins resulting from the inhibition of alpha-chymotrypsin with 5 analoges of diisopropyl fluorophosphate (DFP) showed that in each case approximately 1 mole of phosphorus had been introduced per mole of enzyme in spite of great differences in the amount of the analoge needed to cause complete inhibition. The analoges used were diphenyl chlorophosphate, diethyl thionofluorophosphate, tetraisopropyl pyrophosphate, diethyl p-nitrophenyl phosphate, and tetrapropyl dithionopyrophosphate.

PROBABLE MAXIMUM SPECIFIC ACTIVITY OF MACERANS AMYLASE. S. Schwimmer and A. K. Balls, Fed. Proc. 10(1)pt. I, p. 245, March, 1951. The specific activities were found to be close to those predicted from solubility measurements of impure preparations and were of the same order of magnitude as observed for other pure enzymes.

PHOSPHATE UPTAKE BY PEA MEAL EXTRACTS (with Calif. Inst. Technol.), B. Axelrod, R. S. Bandurski, and P. Saltman, Fed. Proc. 10(1)pt. 1, 158-59, March, 1951. A particulate-free extract of pea meal was found to esterify inorganic phosphate in absence of detectable starch. Similar behavior was shown by soybean meal which is naturally free of starch. The reaction mixture was incubated with radio-active phosphate and products resulting from esterification of the endogenous substrate were isolated by use of two-dimensional chromatography.

A CRYSTALLINE ACTIVE OXIDATION PRODUCT OF ALPHA-CHYMOTRYPSIN. E. F. Jansen, M. D.-F. Nutting, A. L. Curl, and A. K. Balls, Fed. Proc. 9(1)pt. 1, pp. 186-7, March, 1950; Jour. Biol. Chem. 189(2):671-82, April, 1951. Alpha-chymotrypsin when oxidized with sodium periodate in the molar ratio of 5 to 20 moles of periodate per mole of enzyme resulted in formation of an altered enzyme which was isolated in crystalline form. Non-protein nitrogen was formed during the oxidation. The crystalline form of the oxidized enzyme was similar to that of the parent enzyme crystallized under similar conditions.

THE NATURE AND PREPARATION OF ENZYMES. A. L. Curl and S. Schwimmer, Crops in Peace and War, Yearbook of Agriculture, 1950-51, pp. 69-75 (U.S. Dept. Agr.). A general discussion of the nature of enzymes and of methods for their preparation. Among topics discussed are: origin; catalytic action; protein nature; relation to vitamins; preparation and use of commercial enzyme products such as pancreatin, pepsin, papain, fucin, rennin and barley malt; methods of preparation of pure enzymes; outline of preparation of crystalline chymotrypsin and alpha-amylase.

ISOLATION OF AN ORGANIC SULPHUR COMPOUND FROM ASPARAGUS. Patent No. 2,559,625 to E. F. Jansen, July 10, 1951. Asparagus juice is extracted in a series of operations with butanol and benzene to isolate a yellow amorphous material containing about 30 percent sulphur. On reduction this material yields 3,3'-dithiolisobutyric acid which is useful as a regulator for enzymic reactions.

DITHIANE CARBOXYLIC ACIDS. Patent No. 2,559,626 to E. F. Jansen, July 10, 1951. An aldehyde or ketone is reacted with 3,3'-dithiolisobutyric acid to produce dithiane carboxylic acids such as 1,3-dithiane-5-carboxylic acid.

THE FREE ENERGY OF HYDROLYSIS OF p-NITROPHENYL PHOSPHATE. B. Axelrod, Science 114 (2968):525-26, Nov., 1951. The free energy of hydrolysis of p-nitrophenyl phosphate has been determined.

THE CHROMATOGRAPHIC IDENTIFICATION OF SOME BIOLOGICALLY IMPORTANT PHOSPHATE ESTERS. R. S. Bandurski and B. Axelrod (with Calif. Institute of Technol.), Jour. Biol. Chem. 193(1):405-10, Nov., 1951. A two-dimensional chromatographic method for identification of some phosphate esters of biological importance, including inorganic polyphosphates, is described. An acid solvent consisting of methanol, formic acid, and water and a basic solvent containing methanol, ammonia, and water are used. An improved method of color development involving ultraviolet light permits the hydrolysis of resistant esters and minimization of background color.

BITTER PRINCIPLES IN CITRUS. III. SOME REACTIONS OF LIMONIN. O. H. Emerson, Jour. Amer. Chem. Soc. 74, 688-93, Feb., 1952. The hydrogenation of limonin ($C_{26}H_{30}O_8$) gives a mixture of tetrahydrolimonin ($C_{26}H_{34}O_8$) and hexahydrolimoninic acid ($C_{26}H_{36}O_8$). Limonin appears to be bicarbocyclic and has two ethylenic links, one of which forms an allylic system with the potential hydroxyl of a lactone group.

The oxidation of limonin with alkaline hypoiodite gives limonilic acid ($C_{26}H_{30}O_8$). Tetrahydrolimonin and hexahydrolimonic acid behave similarly. The reaction apparently involves the opening of a lactone ring and formation of a new carbocyclic ring. One lactone ring is opened by hydrogenolysis and the other by oxidation. Treatment of limonin with hydriodic acid gives citrolin ($C_{26}H_{22}O_6$) which appears to be tricarbocyclic, and has four ethylenic links, one of which is conjugate with the carbonyl group, and one with a lactone group.

ENZYMIC AND CHEMICAL HOMOGENEITY OF SCHARDINGER DEXTRINOGENASE. S. Schwimmer, Fed. Proc. 11(1):pt. I, pp. 283-84, March, 1952. Further experimental evidence obtained with a purified preparation of the enzyme showed that at least 90 percent of the protein was present as one enzymically active component.

FURTHER STUDIES ON THE PRODUCTION, PURIFICATION, AND PROPERTIES OF THE SCHARDINGER DEXTRINOGENASE OF BACILLUS MACERANS. S. Schwimmer and J. A. Garibaldi, Cereal Chem. 29(2):108-22, March, 1952. A fermentation method which increases the production of Schardinger dextrinogenase by Bacillus macerans is reported. Purified preparations obtained from the culture fluid were studied.

INHIBITION OF BETA- AND GAMMA-CHYMOTRYPSIN AND TRYPSIN BY DIISOPROPYL FLUOROPHOSPHATE. E. F. Jansen and A. K. Balls, Jour. Biol. Chem. 194(2):721-27, Feb., 1952 Diisopropyl fluorophosphate inhibited beta- and gamma-chymotrypsin in essentially the same manner as it inhibits alpha-chymotrypsin (previously reported). Trypsin required greater concentration of DFP for inhibition than the chymotrypsins did. Approximately one mole of phosphorus was introduced per mole of enzyme in the inhibition reactions.

PROTEOLYTIC ENZYMES. A. K. Balls and E. F. Jansen, Ann. Rev. Biochem. Vol. 21, pp. 1-28, 1952. This chapter critically reviews the literature on proteinases for the years 1950 and 1951.

POLYMETHYLGALACTURONASE, AN ENZYME CAUSING THE GLYCOSIDIC HYDROLYSIS OF ESTERIFIED PECTIC SUBSTANCES. C. G. Seegmiller and E. F. Jansen, Jour. Biol. Chem. 195(1): 327-36, March, 1952. From a commercial enzyme product there has been extracted a pectic enzyme, a polymethylgalacturonase, which attacks the glycosidic polyuronide bonds of pectin from citrus, raspberry, and apple. It attacks pectin more rapidly than it attacks pectic acid, has a pH optimum around 6, and is unable to produce galacturonic acid (or its methyl ester) or to hydrolyze more than 26 percent of the available uronide bonds of pectin or more than 17 percent of those of citrus methylglycoside of polygalacturonic methyl ester.

ISOLATION OF ALPHA-AMYLASE FROM MALT EXTRACT. Patent 2,594,356, to S. Schwimmer and A. K. Balls, Apr. 29, 1952. Malt extract is heated and filtered to remove beta-amylase. Alpha-amylase is isolated from the filtrate by a procedure involving salt precipitation, adsorption in a solid amylaceous substance and elution with water. Alpha-amylase is crystallized from the eluate in pure and active form.

INHIBITION OF PURIFIED, HUMAN PLASMA CHOLINESTERASE WITH DIISOPROPYL FLUOROPHOSPHATE E. F. Jansen, R. Jang, and A. K. Balls, Jour. Biol. Chem. 196(1):247-53, May, 1952. When a preparation of purified, human plasma cholinesterase was inhibited by radioactive DFP, the phosphorus of the inhibitor was found to have been introduced into the inhibited enzyme. Hence the inhibition reaction of this enzyme by DFP was similar in this respect to the inhibition of the esterolytic proteinases.

PHARMACOLOGICAL INVESTIGATIONS (July 1, 1950 to June 30, 1952). (A separate list, AIC-220, includes abstracts of publications on pharmacology up to June 30, 1950, and is available on request.)

TOXICOLOGICAL STUDIES ON COMPOUNDS INVESTIGATED FOR USE AS INHIBITORS OF BIOLOGICAL PROCESSES. I. TOXICITY OF VINYL PROPIONATE, A. M. Ambrose, Arch. Indus. Hygiene and Occup. Med. 2(5):582-90, Nov., 1950. Studies on the toxicity of vinyl propionate have been made on experimental animals by various routes of administration. Health hazards are tentatively evaluated.

TOXICOLOGICAL STUDIES OF COMPOUNDS INVESTIGATED FOR USE AS INHIBITORS OF BIOLOGICAL PROCESSES. II. TOXICITY OF ETHYLENE CHLOROHYDRIN, A. M. Ambrose, Arch. Indus. Hygiene and Occup. Med. 2(5):591-97, Nov., 1950. Studies on the toxicity of ethylene chlorohydrin have been made on experimental animals by various routes of administration. Health hazards are tentatively evaluated.

TOXICOLOGICAL STUDIES OF COMPOUNDS INVESTIGATED FOR USE AS INHIBITORS OF BIOLOGIC PROCESSES. III. TOXICITY OF PROPYLENE GLYCOL DIPROPIONATE, A. M. Ambrose, Arch. Indus. Hygiene and Occup. Med. 3(1):48-51, Jan., 1951. Studies on the toxicity of propylene glycol dipropionate have been made on experimental animals by various routes of administration. Health hazards are tentatively evaluated.

TOXICOLOGICAL STUDIES OF COMPOUNDS INVESTIGATED FOR USE AS INHIBITORS OF BIOLOGIC PROCESSES. IV. TOXICITY OF 1,3-DIMETHYL-4,6-BIS(CHLOROMETHYL)BENZENE, A. M. Ambrose, Arch. Indus. Hygiene and Occup. Med. 3(1):52-56, Jan., 1951. Studies on the toxicity of 1,3-dimethyl-4-6-bis(chloromethyl)benzene have been made on experimental animals by various routes of administration. Health hazards are tentatively evaluated.

COMPARATIVE TOXICITY OF PYRETHRINS AND ALLETHRIN. A. M. Ambrose and D. J. Robbins, Fed. Amer. Socs. Expt. Biol. Proc. 10(No. 1, Pt. 1):276, March, 1951. Studies with rats, in which pyrethrin and synthetic allethrin were administered topically, subcutaneously, and gastrically, revealed no toxic effects with purified samples. With less pure samples, slight toxic effects, such as transitory skin irritation and slight decrease in growth rate at high dietary level, were observed. Thus it was concluded that purified allethrin is no more toxic than purified pyrethrins.

STUDIES ON THE CHRONIC ORAL TOXICITY OF COTTONSEED MEAL AND COTTONSEED PIGMENT GLANDS. A. M. Ambrose and D. J. Robbins, Jour. Nutrition 43(3):357-70, March, 1951. From studies with rats on depigmented cottonseed meal, hexane-extracted cottonseed meal, and cottonseed pigment glands, it is concluded that depigmented cottonseed meal is devoid of any toxic principle, and that the toxicity of the pigment glands may be attributable to some constituent(s) of the glands other than gossypol.

COMPARATIVE STUDIES ON TOXICITY OF A NEW STREPTOMYCIN AND STREPTOMYCIN SULFATE. A. M. Ambrose, Proc. Soc. Expt. Biol. and Med. 76(3):466, March, 1951. Hydroxystreptomycin trihydrochloride when tested in mice by subcutaneous administration was found to have about the same order of toxicity as streptomycin sulfate.

PROTECTION BY FLAVONOIDS AGAINST HISTAMINE SHOCK. R. H. Wilson, A. N. Booth, and F. DeEds, Proc. Soc. Expt. Biol. and Med. 76(3):540-42, March, 1951. Slight protection of guinea pigs treated with rutin and subjected to an LD_{50} dose of histamine, previously reported, has been confirmed. Quercitrin, quercetin, and methyl hesperidin chalcone also were found to afford protection against histamine. Protection is slight and is of theoretical rather than clinical interest.

ACUTE AND CHRONIC TOXICITY OF POTASSIUM ACID SACCHARATE. A. M. Ambrose, Jour. Amer. Pharm. Assoc., Sci. Ed. 40(6):277-79, June, 1951. The acute oral or intraperitoneal toxicity of potassium acid saccharate is extremely low. No toxic effects or fatalities were observed in rats receiving as much as 2 g./kg. in a single dose of a solution of potassium acid saccharate previously neutralized to reduce irritation. Rats fed diets containing 2 percent or less potassium acid saccharate for approximately 210 days showed no untoward effect, as judged by appearance, growth, food consumed, mortality, organ weights, and microscopical examination of visceral organs.

PHYSIOLOGICAL EFFECTS OF DISTILLERS' DRIED GRAINS DERIVED FROM ACID-SACCHARIFIED CORN MASH. A. M. Ambrose, Cereal Chem. 28(4):334-39, July, 1951. In studies with rats it was concluded that the toxicity (mild laxative action) of acid saccharified corn mash is due to the presence of calcium and sodium sulfate arising from neutralization of the sulfuric acid used in the saccharification process.

CHRONIC TOXICITY OF LAURYL GALLATE. S. C. Allen and F. DeEds, Jour. Amer. Oil Chem. Soc. 28(7):304-306, July, 1951. In studies with rats there was no evidence that ingestion of slight amounts of lauryl gallate, such as might be encountered in edible fats and oils protected against oxidation, would produce harmful effects. At 1.0 percent level there was some retardation or inhibition of growth. Higher levels used caused no histopathological changes in organs but were obviously a factor in starvation of test rats.

ISOLATION OF QUERCITRIN AND QUERCETIN FROM LEMON FLAVINE. A. N. Booth and F. DeEds, Jour. Amer. Pharm. Assoc., Sci. Ed. 40(8):384-85, Aug., 1951. A simple method for the isolation of pure quercitrin and quercetin from lemon flavine in high yield is described. Application of the method to several samples of lemon flavine has given uniformly satisfactory results.

TOXICITY STUDIES ON SUBTILIN. R. H. Wilson, F. DeEds, and L. J. Rather, Proc. Soc. Expt. Biol. and Med. 78(3):517, 519, Dec., 1951. Subtilin in diets of rats (at levels as high as 0.1 percent for 220 days) was not detrimental as judged by growth, activity, organ weights, and macroscopic and microscopic appearance of the tissues.

COMPARATIVE TOXICITIES OF QUERCETIN AND QUERCITRIN. A. M. Ambrose, D. J. Robbins, and F. DeEds, Jour. Amer. Pharm. Assoc., Sci. Ed. 41(3):119-22, March, 1952. Data are given on possible toxicity of quercetin and quercitrin after intravenous administration in rabbits and in rats and after oral ingestion of diets containing as much as 1 percent of each for approximately 410 days. Results indicate a low order of toxicity. It is doubtful that these compounds exert any specific toxic effects.

INHIBITORY EFFECTS OF ITACONIC ACID IN VITRO AND IN VIVO. A. N. Booth, J. Taylor, R. H. Wilson, and F. DeEds. Jour. Biol. Chem. 195(2):697-702, April, 1952. Rats fed diets containing 1 or 2 percent of itaconic acid experienced a statistically significant inhibition of growth but showed no other toxic effects. Apparently impaired use of succinic acid caused the decreased growth rate.

HAZARDS AND POTENTIAL DRUGS. F. DeEds, R. H. Wilson, and A. M. Ambrose, U.S.D.A. Yearbook of Agriculture, Crops in Peace and War, 1950-51, pages 721-26. A review of the general purposes and problems of the testing of pesticides, insecticides and related compounds for toxicity. Two compounds (phenothiazine and 2-acetaminofluorene) are considered at some length for purposes of illustration.

MECHANISM OF RUTIN ACTION. F. DeEds, U.S.D.A. Yearbook of Agriculture, Crops In Peace and War, 1950-51, pages 746-50. Rational use of a drug requires an understanding of the way it acts in the body. This review of studies on the mechanism of action of rutin discusses experiments that indicate that the antioxidant properties of rutin may be the reason for such beneficial actions as decrease in destruction of epinephrine (with decrease in petechial bleeding) and decreased destruction of ascorbic acid with advantage.

NATURAL RUBBER EXTRACTION AND PROCESSING, 1636 East Alisal Street, Salinas, Calif. (January 1, 1950-June 30, 1952). (A separate list, AIC-276, includes abstracts of publications and patents resulting from previous studies in this Bureau of extraction of rubber from native plants. The investigations covered in the list below were conducted at the Station at Salinas.)

A LOW-MOLECULAR-WEIGHT FRACTION OF GUAYULE RUBBER. J. W. Meeks, T. F. Banigan, Jr., and R. W. Planck, India Rubber World 122(3):301-4, June, 1950. An acetone-soluble fraction was obtained from guayule tissue and from crude guayule rubber by fractional precipitation with alcohol. This bears a considerable chemical and physical similarity to rubber hydrocarbon. Its molecular weight is estimated to be 20,000 or less. Whether this material is present as such in the live shrub where it may play a role in the synthesis of rubber or whether it should be regarded as some degradation product of the higher polyprenes has not been determined.

FIXTURES AID IN SHARPENING DIES USED TO CUT RUBBER TENSILE TEST SPECIMENS. R. H. Taylor, India Rubber World 123(4):441, 449, Jan., 1951. Mechanical fixtures designed to position dies so that the surface to be ground may be firmly held in either a horizontal or vertical plane have been found to aid materially in maintaining sharp dies. Use is made of a high-speed drill press, a small high-speed flexible shaft grinder, an illuminated magnifier, and an assortment of fine stones for sharpening.

DERESINATION AS A MEANS OF IMPROVING THE QUALITY OF GUAYULE RUBBER. I. SHRUB DERESINATION, R. L. Chubb, E. C. Taylor, and I. C. Feustel, India Rubber World 123(5):557-62, 569, Feb., 1951. The low quality of crude resinous guayule rubber is largely ascribed to its resinous constituents. A crude rubber containing resin within a range of about 0.7 to 2.0 percent, in contrast to the normal content of 20 to 25 percent, can be obtained by pebble milling guayule shrub previously extracted with acetone. Freshly harvested shrub is first defoliated, chopped, and crushed. The moist material is then extracted directly with acetone by laboratory and semi-pilot-plant-scale batch countercurrent procedures. Results of physical tests on the rubber obtained from deresinated shrub indicate that it is similar to Hevea (plantation) rubber.

GUAYULE AN AMERICAN SOURCE OF RUBBER. K. W. Taylor, Econ. Bot. 5(3):255-73, July-Sept., 1951. The growth, culture, and early commercial development of guayule are reviewed. Objectives of current research include increasing the productive capacity of the plant through genetic and plant breeding research; determination of most favorable areas for producing rubber; maintenance of an adequate seed supply for emergency planting and for multiplication of new strains or varieties; development through laboratory and pilot-plant investigations of improved or new processes and equipment for economical production of a high-quality and stable crude rubber; and investigation of possible uses for by-products of rubber extraction. Guayule may have possibilities as an economic farm crop in certain marginal dry-land areas of Texas.

TEMPERATURE MEASUREMENTS IN THE MOONEY VISCOMETER. R. H. Taylor and W. P. Ball, Amer. Soc. for Testing Materials Bull. 176, 60-66, Sept., 1951. One of the chief difficulties in obtaining the proper cure curve for a rubber stock in the Mooney viscometer is due to the inadequacy of methods available for control and measurement of temperature of the test specimen. The reliability of various methods is discussed and the relationships are shown between the temperatures indicated by a thermocouple coiled in the test specimen, and temperatures in the platens and dies. Temperature control will probably not be adequately solved until the non-symmetrical characteristics of the machine are corrected.

THE MOLECULAR WEIGHT OF GUAYULE RUBBER AS INFLUENCED BY EXTRACTION AND PROCESSING TREATMENTS. J. W. Meeks and I. C. Feustel, India Rubber World 125(2): 187-90, Nov., 1951. Inherent viscosities of benzene-rubber solutions are used to measure the relative changes in molecular weight of guayule rubber that occur in the several steps of the pebble milling process of recovery and during subsequent storage. A progressive degradation is observed. Drying the rubber with heat is particularly injurious. Deresinated rubber is more stable toward heat than the resinous product. Application of antioxidant offers at least partial protection against effects of heat and storage.

DISTRIBUTION OF WAXES IN GUAYULE. T. F. Banigan, Jr., J. W. Meeks, and R. W. Planck, Bot. Gazette 112(6):231-4, Dec., 1951. Wax extracted from guayule shrub by refluxing the comminuted plant tissues with ethanol was separated from the cooled extract by filtration. The wax was found to occur almost exclusively in the phloem above the crown and in the actively growing parts of the plant. Wax was also found in resinous crude guayule rubber to the extent of at least 0.5 percent. Distribution appears largely independent of season in shrubs of the same age and variety.

PROCESS OF ISOLATING BETAINE FROM GUAYULE EXTRACT. Patent No. 2,549,763 to T. F. Banigan, Jr., J. W. Meeks, and R. W. Planck, April 24, 1951. The process of isolating betaine as its phosphate from the solvent extract obtained in connection with the preparation of resin-free guayule rubber which comprises concentrating the extract, allowing the concentrated extract to separate into a non-resinous phase containing betaine and a resin phase, separating the non-resinous phase, mixing it with an oxygenated organic solvent and acidifying it with phosphoric acid, then separating the betaine phosphate which crystallizes out of solution.

PROCESS FOR RECOVERING PARTHENYL CINNAMATE AND ESSENTIAL OILS FROM GUAYULE RESIN. Patent No. 2,572,046 to J. W. Meeks, T. F. Banigan, Jr., and R. W. Planck, October 23, 1951. A process for isolating parthenyl cinnamate from guayule resin which comprises mixing a water-miscible organic solvent solution, such as an ethanol extract of the guayule resin with a hydrocarbon, such as pentane in which the parthenyl cinnamate is more soluble than in the water-miscible organic solvent, and separating the hydrocarbon containing the parthenyl cinnamate from the mixture.

DETERMINATION OF RUBBER HYDROCARBON BY A MODIFIED BROMINATION METHOD. W. J. Gowans and F. E. Clark, Analyt. Chem. 24(3):529-33, March, 1952. Accurate and precise results can be obtained in the determination of rubber hydrocarbon by the bromination method if adherence to the recommended procedure is observed. The bromination reaction is sensitive to rubber hydrocarbon concentration, time of bromination, temperature, and light conditions. Addition of chloroform to the brominating solution inhibits substitution and brings the empirical factor for conversion of rubber bromide to rubber hydrocarbon close to the theoretical factor for a completely additive product.

RUBBER RECOVERY FROM FRESHLY HARVESTED GUAYULE. K. W. Taylor and R. L. Chubb, Indus. and Engin. Chem. 44(4):879-82, April, 1952. Freshly harvested guayule shrub can be processed successfully for the recovery of rubber without the addition of coagulant chemicals. Coagulation of the latex in fresh guayule may be brought about by parboiling and mechanical treatment, though the most efficient method for doing so remains to be developed. Shrub conditioned and stored in the conventional manner associated with commercial operations gives decreasing rubber hydrocarbon yields with increased periods of storage. Rubber from shrub handled in this manner exhibits progressive deterioration.

DEVELOPMENT OF PILOT PLANT CONTROL FOR MILLING GUAYULE SHRUB. K. W. Taylor and R. L. Chubb, Indus. and Engin. Chem. 44(4):883-87, April, 1952. A standardized pebble milling control procedure was developed by use of biological statistical design and analysis. Such designs and analyses showed that factors heretofore unknown or thought to be insignificant were of major importance. This procedure provides criteria for judging the effectiveness of changes in methods or equipment in the development of new or improved processes for large-scale production of guayule rubber. The same principles are applicable to other fields of process research.

CPSIA information can be obtained
at www.ICGtesting.com
Printed in the USA
LVHW080808051218
599325LV00003B/262/P